SpringerBriefs in Business Process Management

Series editor

Jan vom Brocke, Vaduz, Liechtenstein

For further volumes:
http://www.springer.com/series/13170

Joachim Van den Bergh
Sara Thijs · Stijn Viaene

Transforming Through Processes

Leading Voices on BPM, People
and Technology

 Springer

Joachim Van den Bergh
Technology and Operations Management
Vlerick Business School
Ghent
Belgium

Stijn Viaene
Vlerick Business School
Leuven
Belgium

Sara Thijs
Technology and Operations Management
Vlerick Business School
Leuven
Belgium

ISSN 2197-9618   ISSN 2197-9626 (electronic)
ISBN 978-3-319-03936-7   ISBN 978-3-319-03937-4   (eBook)
DOI 10.1007/978-3-319-03937-4
Springer Cham Heidelberg New York Dordrecht London

Library of Congress Control Number: 2013957705

Printed on acid-free paper

Springer is part of Springer Science+Business Media (www.springer.com)

Preface

Business Transformation is the talk of the town. Organisations are now facing challenges that are questioning their very reason for existence. Who is our customer, how to serve them, which processes should we install, how do we approach the market, which technological trends could disrupt our sector, ... ? All these are valid questions that need to be answered on a continuous basis if organisations want to stay relevant and at least in tune with the competition.

Now the question that arises is whether BPM will be able to maintain its position as an established management discipline in this type of environment? Originally, BPM was introduced as an instrument for continuous improvement, automation and standardisation. The contemporary business environment, in contrast, calls for resilience, agility and innovative capacity. So we should ask ourselves as BPM academics and practitioners: Is BPM here to stay, and what does it take?

Instead of trying to answer the question ourselves we consulted international experts to consider the challenges for BPM, its future and how it is applied under different circumstances. That is why we present to you this series of 14 short interviews with inspiring practitioners and academics sharing insights into the wonderful world of process and transformation. We are especially glad to have talked to the HR officer of an SME and the head of a governmental agency's Process Innovation Team, to report on areas of application that are largely neglected by researchers and that have their own interpretation of 'process thinking'. With this publication, we wanted to go beyond the typical 'how to implement BPM' stories, and instead approach BPM from various less familiar perspectives such as culture, the employee and customer points of view, and how BPM is perceived in a fast developing BRIC country, namely Brazil. The interviews with the academics are focused on the cross-over of BPM and other knowledge areas such as Business Intelligence, emerging technologies and innovation management. Without being exhaustive, we are confident that these insights will prove both engaging and useful for the reader of this document.

Having studied the field of BPM for some time now at Vlerick Business School, we felt that it was time to help draft the agenda for another era of transformational thinking. We hope that by picking up this interview book, whether you are a researcher or a practitioner, you will join us in our attempts to achieve the next level of business process excellence.

Enjoy the read and be inspired!

Joachim Van den Bergh
Sara Thijs
Stijn Viaene

Contents

Abstract

The management discipline business process management (BPM) has come of age as a mature instrument for organisations to install a methodological approach for modelling, analysing, improving and streamlining processes. Yet, practitioners and academics feel that BPM has perhaps not lived up to the high expectations and still poses a lot of implementation challenges.

Meanwhile we have arrived in a new transformational business era, characterized by digital innovations, economic turbulence, transforming businesses and extreme customer mobility. Are we ready to support our organisations with process thinking in this environment? In 14 interviews with practitioners and academics we have attempted to answer what BPM could look like in the near future, how different organisations apply it and which challenges remain to be solved.

Using New Digital Technologies to Innovate Business Processes and Create Customer Value: An Interview with Prof. Stijn Viaene

New digital technologies are penetrating every aspect of our personal and professional lives—and organisations are under pressure to adapt. To gain greater insight into how these new technologies are affecting the field of BPM, we asked Prof Stijn Viaene for his views on the matter.

For a number of years now, Stijn Viaene has been passionate about the relationship between business and IT. His research and teaching centre on how to create value with IT in the context of business process improvement. He points out that, despite all the buzz around new technologies, the traditional definition of a business process has remained a stable one: a business process can still be defined as *a collection of activities that together create value for a customer*. Moreover, improving your organisation by viewing it through a process perspective has not lost a bit of its value.

Stijn Viaene: "The big change is in how digital technologies allow us to re-conceive customer value creation by innovating business processes with information and information exchange. Traditionally, organisations have focused on the 'what' of processes in BPM exercises, and technology has often been brought in merely for automation. Now, new digital technologies are enabling organisations to escape from thinking solely in terms of automation as a solution and to start thinking about making processes smarter as well as more efficient."

According to Prof Viaene, the technologies that are most impactful are those that support advanced analytics and big data, social media, and new means of IT solution delivery such as mobile and cloud computing. Each technology certainly has its own contribution to make—but, taken together, they create a most powerful platform for change.

Linking Up with Customers' Personal Lives

Take big data. Today, behaviour, preferences and intentions are constantly being tracked while a customer lives his/her online life. This potentially allows better understanding of, and tremendously richer, interactions with customers. Real-time action based on this data is

J. Van den Bergh et al., *Transforming Through Processes*, SpringerBriefs in Business Process Management, DOI: 10.1007/978-3-319-03937-4_1, © The Author(s) 2014

already feasible – and many customers now expect it to be used to their advantage. But many of our organisational processes are having trouble accommodating to this new context; in fact, processes will have to be seriously revisited to take advantage of this opportunity. To meet the here-and-now needs, websites should be re-designed on the fly, based on a real-time customer analysis. Market trends need to be mined from technologies such as Twitter – which its CEO calls 'a modern-day agora'. Also, intentions are to be gleaned from analysing personal search history using tools like Google.

With social media and mobile data at their fingertips, organisations can more easily link up with their customers' personal lives. Marketers will no longer have to rely on coarse-grained profiling methods—or, as is often the case, guesswork—to determine how to engage with their customers. Prof Viaene puts it powerfully: "Social media and big data analysis enable organisations to get significantly closer to turning the customer's real-time process into their business process."

Certainly, cloud computing enables more straightforward sourcing of business processes: it allows organisations to reconfigure their operating models by focusing on their own choice of core capabilities and transferring the responsibility of providing other processes to partners. Yet, Prof Viaene adds a caution: "The consumerisation of digital technologies has dramatically increased the risk of ending up in more silos than ever before – cutting up organisations, their business processes and, most importantly, the customer experience, into thousands of little pieces. Enterprise-wide management and governance of IT solutions and platforms will remain crucial to survival."

Re-Balancing

"Currently, customer-facing processes seem to be getting most of the attention in terms of the potential inherent in using digital technologies to revisit business and business models—pulling customers into value co-creation scenarios, for example." The HR domain has also received quite a lot of attention: e.g. stimulating self-selection of work in crowd-sourcing approaches to staffing. But there are so many other areas that can benefit. Entire value chains are on the verge of being reengineered. This is because we started fixing digital sensors to every person (e.g. mobile devices), every product (e.g. RFID), every machine… in short, every 'thing'. We're getting ever closer to having a 'web of sensors' at our disposal—also called an 'internet of things'—to rethink business engagement. This will undoubtedly revolutionise the way businesses will be organised and engage with human beings in the process of value creation. This generation of 'big data' will come with big opportunities as well as big challenges—for example, the social and ethical issues of privacy and security.

So, we absolutely need to step away from a reductionist view of technology that aims only for cost-cutting and more process throughput. We can, and should, get far more out of new digital technologies. Processes for value creation can be made more intelligent, more collaborative – truly social. In a nutshell, we can make business quite a bit 'smarter'.

Does that mean that we need to replace the focus on standardising processes and IT with a focus on customising the use of technology? "Not exactly – but we need to re-balance things. In the past, companies have often over-emphasised standardising the use of IT to achieve economies of scale by creating lean shared IT service centres or outsourcing. In many cases, this scenario can bring quite a few benefits, but it is not the only, nor the most desirable, scenario in terms of appropriate use of IT. Many activities that organisations engage in can and should be standardised – but some things should simply not be standardised: 20–30 % of a business's processes truly distinguish a customer value proposition. It's not a question of substituting but balancing, with the right scenario chosen for the right reasons, taking your strategy, your customers and the context in which you operate into account."

Adopting Technology Mindfully

Will new digital technologies ensure competitive advantage? "Value creation with IT is not, and has never been, about the IT resource per se. We have to consider the whole organisational ecosystem and how to align the technology, the processes and the people. We have to master the way these parts interact in order to create value."

> Granted, there's a lot of hype around these technologies. But that shouldn't stop us from experimenting today and learning how to deploy them for value creation. Some organisations seem to have settled for 'technology watch' mode. But that's not a wise move. It's time to experiment, feel and see what is possible and be part of 'making things happen'. If you don't pick up the glove now, someone else will.

> Companies can leave their competition way behind by adopting technology purposefully—i.e. to support a strategy and customer value proposition—as well as mindfully— i.e. considering its end-to-end impact on their business ecosystem.

Stijn Viaene is Partner at Vlerick Business School, where he heads the Technology and Operations Management (TOM) area. He is a tenured Professor in the Decision Sciences and Information Management department at KU Leuven. Stijn's research and teaching focus on information systems management (ISM) issues in three primary areas: (1) Business-IT alignment, strategy, leadership and governance; (2) Business process management (BPM); and (3) Business intelligence (BI).

A Government Institution's BPM Tale: An Interview with Bert Schelfaut of VDAB

Some of society's biggest service providers are government institutions. However, their effectiveness is often hampered by their bureaucratic and cumbersome operations. We visited VDAB, the Flemish public employment service, which has launched initiatives to embrace new technological developments and become more efficient. Bert Schelfaut, Manager of VDAB's Corporate Process Innovation Team, is on the frontline when it comes to guiding the organisation into the future. We asked him for his insights on the role played by BPM.

VDAB was having difficulties translating management's vision to the work floor. So, 3 years ago, BPM was introduced primarily to serve as a link between the two. But VDAB also wanted to implement BPM to make its internal working procedures more transparent—to show where funds are being invested and where VDAB needs to optimise its processes. The organisation created the Process Innovation Team (PIT) to guide the BPM implementation. BPM has been introduced in internal departments, and a pilot exercise is running in the regional offices.

Bert Schelfaut: "Implementing BPM will create a more flexible organisation. The goal is to have an overview of our core processes, to work with them in a uniform manner, and to use BPM as an instrument for managing the organisation."

Gaining Internal Credibility

BPM will not simply 'freeze' existing processes as they are. "On the contrary, gathering 'As-is' information on processes facilitates detecting problematic situations so that you can re-design certain processes and renew the business. But we must be careful not to lose ourselves in detailed descriptions—understanding large building blocks, and how they are related, is essential."

Familiarising the organisation with BPM as a management technique is just one of the team's duties. Obtaining an overview on enterprise architecture and managing and prioritising the project portfolio are additional responsibilities high on

J. Van den Bergh et al., *Transforming Through Processes*, SpringerBriefs in Business Process Management, DOI: 10.1007/978-3-319-03937-4_2, © The Author(s) 2014

the agenda. But carrying out these tasks successfully depends greatly on the cooperation of all parties involved and on their joint collaboration. The PIT is expected to guide and stimulate people and to provide the framework for institutionalising BPM practices. So how do they cope with the high expectations when their role is essentially to sensitise instead of command? "Steadily gaining internal credibility and ensuring strong leadership is vital. To build this solid reputation, we're constantly repeating our vision and the underlying rationale. Gradually, but surely, we're making progress."

Providing Basic Processes

An additional challenge that Bert sees coming up entails a strategic repositioning exercise. VDAB is redefining its traditional position as 'a mechanism aligning supply and demand in the labour market' towards becoming 'a holistic career supporting institution'. Whereas VDAB used to focus on serving job-seeking individuals, it is now striving to involve all citizens actively throughout their entire working life.

Obviously, such a change—which challenges the roles of all stakeholders— impacts existing processes profoundly. The PIT is working hard to define (or re-define) core processes to fit the new organisation. This repositioning also impacts the existing IT portfolio, as new technological developments are being introduced regularly. One example is an online platform where people can post all kinds of information, from diplomas to personal development plans. This platform should facilitate interaction between employers and job-seekers, while also encouraging more personal career management.

Moreover, external developments are challenging the organisation as well. The public is increasingly employing alternate means to obtain desired services instead of following the standard processes provided by VDAB. "I admit it's becoming difficult to control and steer. Consider, for instance, the apps phenomenon. People are using externally developed apps that interfere with VDAB's processes. But as these digital trends are here to stay, it will become VDAB's role to monitor developments as we evolve into the future. Our role will be to provide the basic guiding processes. In a stable and balanced process system, internal processes can freely interact with external processes."

Enabling the Transformation

So, what is the outlook for the PIT? VDAB will always need a governing body to oversee its IT projects. "But we must not become a department, because VDAB certainly doesn't need a new silo. Instead, a small flexible team should take up this role. As for BPM, my team's job is to guide the implementation of BPM as a management technique until it is institutionalised. Ideally, BPM should be steered by the management committee, and then our guidance will become redundant."

In the meantime, the team will continue spreading the BPM message throughout the organisation via proactive communication. Bert Schelfaut acknowledges that, despite its ambition, VDAB is still a long way from being recognised as a holistic career supporting institution. But a tool like BPM can be an invaluable aid in getting there: "Having that helicopter view on the organisation and managing by processes will enable the transformation."

Bert Schelfaut is Manager of Corporate Process Innovation at VDAB. This team was set up in 2009 to facilitate the organisation's path to greater agility, with business-IT alignment as its baseline. Prior to this position, Bert developed and managed the VDAB multi-channel contact centre for 12 years and coordinated a large European software project (2 years).

VDAB is the Flemish Agency for public employment. They are mandated to match supply and demand between open positions offered by employers and job seeking individuals in the Flemish Region in Belgium.

Business Process Improvement: Questioning the Status Quo—An Interview with Michael zur Muehlen

Business process improvement (BPI) is an intrinsic aspect of effective business process management (BPM). And looking for that competitive edge is what drives business process improvement. Yet organisations have a strong tendency to stay in their comfort zone, and they are averse to the organisational change that often accompanies BPI. These conflicting drives are well worth looking into. We asked Prof Michael zur Muehlen of the Stevens Institute of Technology in the USA to share his views on business process improvement.

Michael zur Muehlen: "Engaging in business process improvement is vital! Competitive advantage is fleeting, as competition catches up quickly." To escape this constant competitive war, organisations can attempt to create a more sustainable competitive advantage by looking for less replicable improvements. For example, consider how Google replicated the annual influenza reports from the Center for Disease Control (CDC). When Google used its search engine information to replicate the influenza spread patterns over time, they discovered a one-to-one match with the CDC's data! They then created a predictive capability that enabled them to forecast trends before anyone else could.

> Yet, fleeting competitive advantage is not a main driver in all sectors. In the public sector, organisations strive to fulfil their mission and serve their constituents in the best possible way with limited resources. For instance, in response to a crisis, the Red Cross must allocate resources optimally in unknown territory. If a novel technology is introduced that creates transparency at the local level, the Red Cross will be able to fulfil its mission more effectively. Social media platforms such as Twitter enable these improvements, but many organisations are only beginning to explore integrating these technologies into their business processes.

Driving Business Process Improvement

Who in the organisation should drive business process improvement? "Improvement can be initiated by anyone who questions the status quo of organisational processes. And this should not necessarily be the business process office (BPO). The BPO is first and foremost a service provider that supplies know-how,

J. Van den Bergh et al., *Transforming Through Processes*, SpringerBriefs in Business Process Management, DOI: 10.1007/978-3-319-03937-4_3, © The Author(s) 2014

frameworks and methodological tools that can support improvement. But, because the members of a BPO don't usually work in the trenches or interact with customers on a daily basis, they may not be as frequently confronted with business problems, constraints or opportunities that lead to improvement scenarios. Furthermore, they might not question the existing product portfolio sufficiently. So, the BPO certainly doesn't have to initiate *all* improvement."

Improvement can result from ground-breaking new ideas, but, in fact, this innovative approach is just one of four improvement dimensions Michael presents in his teaching sessions. While innovation is about incorporating experimentation to create a new concept or approach, improvement can also be more incremental and controlled. For example, making certain adjustments within an existing situation can make processes less costly, more responsive, or better suited to clients' needs.

Other improvement tactics include enhancing current practices by systematically applying a catalogue set of patterns, deriving better practices from other businesses, and leveraging excess or freed-up capacity. "Unfortunately, BPM tools do not currently support these improvement tactics. They support mapping processes, and some even support process design – but BPM tools that actively seek improvement patterns in processes have not yet been commercialised. That would require combining process performance data (events relating to activities, frequencies, timing, etc.) with business data. Data-mining techniques can be used to recognise which business data predict a specific process flow. This interplay between business intelligence and workflow is a promising area, but an out-of-the-box solution is not yet available."

Overcoming Hurdles and Pitfalls

Are companies sufficiently engaged in BPI? "The short answer is 'no' – but there are certainly reasons for this. Process improvement is not easy, and each sector or market has its own challenges: monopoly versus fierce competition, profit versus non-profit, production versus service industries – the balance between running the business and the need for improvement differs everywhere. A common hurdle for initiating process improvement is resistance to organisational change. People are just too comfortable with the way things are. Also, a lack of benchmarking might lead people to think that the current way of operating is effective."

Plus, a number of pitfalls can be encountered during BPI itself. "A frequent first source of trouble is skipping the crucial step of *defining the real problem*. Too often, people jump immediately to assumed solutions, and thus they overlook non-evident solutions. I also think that BPI is often too ad-hoc. Nothing beats inspiration, but a more systematic approach can assist in leveraging spontaneous ideas or substitute for a lack of them. Applying creativity techniques, a clear set of solution criteria, and structured methods to evaluate ideas financially and technically are all tools to instigate a more structured process of creativity."

Another common mistake is trying to improve everything at the same time. "It's better to plan stages of improvement, a phased progression. Also, always ensure there is a fall-back position for when new operations don't go as smoothly as planned. Even very small mistakes can lead to huge process backlogs!"

BPM: Controlling Core Processes Effectively

How does Prof zur Muehlen see the future of BPM? "Core processes have been improved intuitively for ages, and this will continue. No matter what catchy acronym BPM software might carry in the future, BPM will always be relevant as a way of managing core processes efficiently so that organisations can play from their strengths. It helps organisations allocate scarce resources and it provides insights into where to differentiate. Look at the airlines: their granular processes are all basically the same, but some – like Singapore Airlines – manage to differentiate themselves by providing an exceptional customer experience. This can only be done when the organisation controls all of its core processes effectively. In the end, BPM doesn't make companies the same – it makes them better!"

Michael zur Muehlen *is Associate Professor of Information Systems at Stevens Institute of Technology in Hoboken, New Jersey, where he directs the Center for Business Process Innovation. Michael's research interests focus on the organisational aspect of BPM technology, risk-aware process management, and process support for managerial decision-making.*

Process Excellence for Sustainable Business Growth: An Interview with Joost Claerbout of Barco

Reliable and efficient business processes are a precondition for sustainable success or growth. The recent story of Barco—investing in process excellence to support ambitious growth objectives in a turbulent environment—proves in practice what theory has prescribed.

Joost Claerbout, Barco's VP Quality and Process Excellence, describes how his team of process consultants is professionalising BPM at this pioneer in visualisation technology.

Barco has a strong sense of pride in product innovation and quality. Yet, in 2008, in the turbulence of the financial crisis, the company encountered a number of setbacks. It became clear that, to ensure sustainable growth, Barco would have to intensify its efforts in streamlining business processes across its various business units.

The new CEO, Eric Van Zele, entrusted Joost Claerbout and his Process Excellence Office (PxO) team with the mission of supporting the professionalisation of BPM at Barco. The team would serve as a facilitator in change and process management initiatives with a focus on product quality. In addition to the PxO, the Quality Management group—covering certification, quality control and compliance—was brought under Joost's direction.

Taking a Cross-Functional View

Barco defines BPM as a way of working together on processes in the most reliable, repetitive and predictable manner by means of clearly defined roles and responsibilities over the company's different regions, functions and business units. Therefore, Joost staffed the team with high potentials sourced throughout the organisation. Today, the team continues to be a melting pot of functional profiles united by the ability to view the company cross-functionally and to approach internal business partners with an attitude of trust. Typically, PxO team members use their function as a steppingstone to other challenges inside or outside Barco.

J. Van den Bergh et al., *Transforming Through Processes*, SpringerBriefs in Business Process Management, DOI: 10.1007/978-3-319-03937-4_4, © The Author(s) 2014

Joost Claerbout: "We do not try to add another idea or opinion to a discussion. Rather we want to facilitate the discussion by creating a consensus. We assume that the business experts have the required knowledge, and that they only need assistance to develop a solution – a new or improved process – in a sustainable and cross-functional way."

Thinking and Acting as an Integrated Entity

The PxO team reviews its mission and raison d'être critically every year. As a team of internal consultants, they work on programmes that support the corporate strategy—and as the strategy evolves, the PxO's programmes evolve.

> The choices made by our leadership, personified by the CEO, are the main driver for selecting our programmes and projects. And in addition to strategy-driven programmes, our team also engages in smaller Lean SixSigma process improvement projects. The current programme – called 'Barco to be one' – expresses the need to think and act as an integrated entity, instead of as a collection of functions and business units. It also underscores Barco's ambition to lead in its target markets.

The programme has produced initiatives to improve channel management, strategic marketing, and Barco's culture and values. Still, business growth is the main strategic objective inspiring these initiatives.

> When I first started here, we relied heavily on external consultants to execute our improvement initiatives. That seemed like an easy solution to getting certain things done. Today, with the 'Barco to be one' programme, business project managers have been engaged to make the business more committed to improvement projects. Business project managers work alongside PxO internal consultants, reporting project progress and results to the company COO and to me. In a second phase business sponsors were activated to monitor and steer improvement projects. The PxO team continues to function as a project office, providing methodological and sourcing support, but now the responsibility resides more with the business.

Just as Much Energy, More Discipline

As the PxO team has focused attention on business process management, the company has transformed in many ways. More functions have been organised on the corporate level—serving all business units—as opposed to separate functional units within each business unit. The internal best practice or 'primus inter pares' is selected as a standard. Furthermore, the ERP backbone system is moving to a single platform.

> Initially, I found a culture at Barco that I would describe as 'high energy, low control'. The various groups were very proud of their product's capabilities and passionate about the freedom they had in creating those products. The PxO team's challenge has been to maintain that passion and energy, while increasing the level of control to improve efficiency. A key success factor for BPM implementation is credibility – and the only way to sell a higher level of control and discipline is to be successful and then to spotlight those successes.

Managing: and Owning—Business Processes

Although process owners have been appointed, Joost acknowledges that cross-functional ownership has not yet been formally established. The organisation still relies on the PxO team to create traction for cross-functional exchange and standardisation. "Nevertheless, I am confident that the work and achievements of the PxO team have had a considerable impact on Barco by enabling sustainable growth. It's hard to overstate the pace of change we are experiencing. Process excellence is a precondition to competing." Not coincidentally, Barco CEO Eric Van Zele mentioned *business process management and ownership* as key success factors upon his election as Manager of the Year 2012 in Belgium. Joost offers a final important observation for BPM practitioners: "Don't build momentum for a non-strategic project – that's a waste of time and resources! Trying to improve everything without strategic prioritisation simply doesn't work, because projects always mean extra work and stress for the organisation. Focus on strategy-infused programmes."

Joost Claerbout is VP Quality and Process Excellence at Barco. His previous experience has included consulting positions at Anderson Consulting and Accenture, through which he developed his expertise in understanding and improving business processes and guiding change projects.

Value-Oriented Process Modelling: An Interview with Jan vom Brocke

Process modelling is a core aspect of BPM. But overemphasis on modelling activities has triggered failure in countless BPM efforts and has given process modelling a negative image. Nonetheless, these models can be of tremendous value when employed correctly. We had an inspiring talk with Prof. Jan vom Brocke, Director of the Institute of Information Systems at the University of Liechtenstein, about leveraging process modelling in BPM.

What can be done about process modelling's negative image? Prof. vom Brocke recommends building on BPM's legacy and taking a comprehensive view of BPM in which process modelling is but one aspect.

Jan vom Brocke: "Under no circumstances should we move completely away from BPM and/or process modelling because of previous failures. Trying to reinvent the wheel will only lead to companies making the same mistakes all over again. Strategic alignment, organisational culture, people and skills, governance, IT systems and methods are core factors every organisation should take into account when engaging in business process management. But these six factors are not a cookbook – they're a stimulant for embracing an all-inclusive approach towards BPM. Context sensitivity with regard to the company's current state of affairs is crucial. Each company should adjust its approach and balance the various factors in order to optimise value creation from BPM efforts."

Making the Financial Impact Transparent

One of the many benefits of process modelling is its contribution to assessing the value of a proposed process improvement. Still, many organisations struggle to formalise this value assessment. "One concept I find particularly helpful is called 'value-oriented process modelling'. This novel technique allows companies to convert their process model into a financial plan, which enables them to critically assess the financial impact of implementing the proposed improvement."

With value-oriented process modelling, each potential process improvement can be assessed via flexible spread sheets that indicate ROI, pay-back timeframes,

J. Van den Bergh et al., *Transforming Through Processes*, SpringerBriefs in Business Process Management, DOI: 10.1007/978-3-319-03937-4_5, © The Author(s) 2014

and so on. Both the initial investment to alter an existing process and the cost-benefit effects of running a new process can be simulated. This technique helps process modellers build a business case by making the financial impact of a proposed process change transparent.

Weighing All Factors

However, assessing the value of a process improvement is not solely a matter of financial justification. "You should first prioritise process models based on their strategic value and their impact on all stakeholders – employees, customers, managers, suppliers, etc. And then look at the financial justifications with tools like value-oriented process modelling. Sometimes, potentially high strategic value might outweigh unfavourable financial conditions: for example, in projects where processes are altered to fit better with the organisation's culture. In short, the financial tools should serve as instruments for decision support, but they should never be used as the only criteria for making the decision. Furthermore, organisations should be aware that not all aspects of a process can be quantified. Relying on numbers that do not represent reality can undermine the business case for a process improvement".

Tailoring Process Modelling Methodologies

In practice, Jan sees two ways an organisation can allocate the tasks involved in process modelling. "First, you have a decentralised or coaching model, where processes are modelled by the employees performing them. This approach stimulates organisational understanding of, and a sense of responsibility towards, the processes under study, but it's time-consuming. Admittedly, motivating employees to combine process modelling with their daily tasks is a challenge. The other way – a centralised approach – is where a small group of people model processes and spread the message by means of workshops and change agents. This approach – creating a kind of 'BPM centre of excellence' – is probably more efficient and sustainable in the long run".

How might we further enhance process modelling practices? "We should try to integrate more than financial aspects in modelling tools – the value of a process model can be enriched by tools that assess, for example, an organisation's cultural readiness for BPM. In general, all tools could be made more intuitive, social, visual, and collaborative. Furthermore, I recommend revisiting the rich legacy of process modelling methodologies. Most were developed for modelling structured sequential events, and so they might not be suitable in other contexts, like creative industries, for example. To avoid having businesses give up on process modelling, we might want to invest in exploring and conceptualising modelling methodologies tailored to specific business needs".

Finally, Jan stresses that companies must not overemphasise process modelling. "All these recommendations are useful only when companies ground process modelling in a more comprehensive view of BPM. We must not lose our BPM heritage, because it will always be about understanding, improving and managing business processes".

Jan vom Brocke is the Hilti Chair of Business Process Management and Director of the Institute of Information Systems at the University of Liechtenstein. Jan has more than 15 years of experience in IT and BPM projects, and he has published more than 200 peer-reviewed papers in renowned outslets (including MIS Quarterly). Jan has authored and edited 17 books, including: Springer's "International Handbook on BPM" and "Green BPM – Towards the Sustainable Enterprise". Jan is an invited speaker and trusted advisor to organisations around the world.

Integrating Business Intelligence into Your Business Processes: An Interview with Öykü Işik

Is information a company's most important asset? Vlerick Prof Öykü Işik urges us to look more closely: "Information without the skill set, brains and infrastructure to process it just doesn't make sense".

And this is exactly where Business Intelligence (BI) and BPM come into play. Organisations need both the analytical capabilities and the business processes to make sense of information. But because both BI and BPM are organic concepts that change along with the technology that enables them, the challenges associated with them evolve as well. We asked Prof Işik to provide her insights into the most urgent challenges and how organisations can meet them.

Are BI and BPM a dream team? Öykü Işik: "Although BI and BPM are still used separately in organisations, a merging of the two concepts is becoming apparent." BI is leaving the ivory tower of IT and travelling down towards the masses. Beyond basic reporting for managers, BI is increasingly being consumed for transactional purposes or process improvements. Meanwhile, an opposite trend is taking place in the BPM camp. Initially, BPM was a means of building more cost-effective processes. We've observed a bottom-up journey, as now BPM is being depicted as a holistic management discipline.

> Eventually, for the two concepts to fuse, they will have to meet somewhere in the middle so that analytical capabilities can be applied at the business process level. Intelligent processes, process intelligence, process discovery, automated process mining, individualised or customised processes – these are but a few examples of how analytical capabilities integrated into processes can be leveraged. It's all about adapting processes according to the data organisations are collecting.

Big Data is Growing Bigger and Faster

Yet, due to changing dynamics, Öykü foresees several challenges to leveraging this interplay to the fullest. In addition to the unpredictability caused by the organic nature of BI and BPM, both the volume and the nature of data consumed

J. Van den Bergh et al., *Transforming Through Processes*, SpringerBriefs in Business Process Management, DOI: 10.1007/978-3-319-03937-4_6, © The Author(s) 2014

for BI purposes are also evolving. "Big data today will not be the big data in 20 years. Even now, online or mobile social transactions alone generate over 2.5 quintillion (10^{18}) bytes of data per day. Most organisations have neither the storage nor the analytical capabilities to process all of that data. Companies are already hiring social media analysts and installing additional hardware and software platforms."

> Another challenge lies in coping with the increasing speed at which real-time data is emerging. Moreover, big data consists of all types and formats, which are most often unstructured. How to process audio files, videos, images, data from sensors, and more... simultaneously? Finally, companies that manage to create value out of big data are currently thought to possess a competitive advantage – but I expect that, like ERP systems, employing big data will become the norm in the near future. How can organisations still stand out when big data becomes a commodity?

From Real-Time to Right-Time Data

Notwithstanding this range of challenges that go with managing big data, Öykü still believes that big data can be of tremendous value for companies. So how can companies prepare themselves for the future? "Get the junk out! Stop collecting all of the data out there – that's no longer feasible. Choose mindfully, and make a business plan for it. Invest time and energy in answering some critical questions. What do we need big data for? What do we want to get out of it? Which data will be useful for which specific purpose? But even when you're not looking to solve a specific problem or question, don't randomly explore all the data in search of patterns. Take time to select and mine data thoughtfully".

> Equally important is to switch from real-time data to right-time data. Avoid devoting effort to gathering all real-time data – get it only when you need it (i.e. at the right time). To summarise briefly: avoid data pollution by installing a filtering mechanism.

Öykü offers in-memory analytics as a creative solution. The idea behind this technology is to stop storing all data in databases. Instead, collect, analyse and process data in parallel, and immediately dump what is not useful.

Absorbing Data Immediately into Processes

After having selected data that provides value, the next question is: are our processes able to make sense of this data? Do we have the right flow and capabilities? "Companies will have to be ever more proactive in digesting data immediately in their processes. Take the concept of complex event processing (CEP), for example. CEP is about analysing processes and process events repeatedly in order to derive patterns or conclusions that help predict what can happen in real-time. Based on these patterns, negative incidents can be predicted and remedies can be applied in advance".

Call centres are a good example of where CEP can improve the relationship between customer and organisation. By repeatedly analysing customer calls, companies can discover patterns in tone and voice that indicate when customers are becoming upset. When employees are made attentive to such signals in real-time, they can act immediately and avoid losing the customer. For instance, they can try to turn the customer's negative experience into a positive one by prioritising the issue or offering a discount.

And it doesn't stop there. Companies can re-design entire processes according to the data they are collecting. For instance, intelligent processes can learn from BI reports and automatically adapt to actual business events. In the future, those companies that have rich enough data will have the opportunity to customise or individualise all processes for the customer. In the end, this all depends on how companies manage their technological resources and organisational capabilities.

Öykü Işik is Assistant Professor at Vlerick Business School. Her main areas of expertise involve business intelligence and business process management. She holds a Ph.D. degree in Business Computer Information Systems from the University of North Texas. During her doctorate, she developed a business intelligence success model. BI is still a major topic in her research.

Adopting the Outside-in View: Customer-Oriented BPM—An Interview with Henri Buenen and Nele Aendekerk (EDF-Luminus)

Most managerial BPM handbooks claim that, ultimately, BPM is in place to create higher customer value. Yet, practice reveals that many companies are struggling to have BPM live up to its promise.

At EDF-Luminus, maintaining and intensifying a customer-oriented approach is what drives the business. Henri Buenen, Director Customer Service, and Nele Aendekerk, Customer Experience Manager, shed light on their journey towards greater customer orientation.

BPM at Luminus serves two purposes: first of all, to simplify the customer process; and secondly, to create more internal efficiency. Henri Buenen strongly believes that there shouldn't be a large trade-off between the two: "Ultimately, a simple and smooth customer process should be supported by efficient internal processes".

'Embracing the customer' is not a new concept at Luminus—excelling in customer service has been one of the company's main objectives since the very beginning. Yet, whereas market forces previously occupied a lot of the company's attention, Luminus has recently renewed its focus on the long-term goal of *customer orientation*.

To achieve this goal, the company has initiated a transformation programme and launched several improvement projects. These initiatives involve redefining customer standards (what does a customer expect from Luminus?), developing a customer process in line with these standards, and adapting the internal business processes accordingly. More than ever, Luminus recognises that customer orientation is about managing customer expectations and delivering high-quality service.

Fundamental Process Changes

The transformation path has entailed going beyond incremental customer process improvements. Henri Buenen and Nele Aendekerk agree that: "For many years, initiatives have led to improvements with marginal value, not really simplifying the customer's experience". Over time, this 'current state +5 % improvement' approach became intolerable. Installing a customer process that is founded on customer standards has required fundamentally redefining core customer processes, and, subsequently, the underlying internal business processes.

A prominent example is the 'move' process, which is currently being piloted. Nele Aendekerk explains: "Even though this process was very well controlled internally, it did not fulfil current customer expectations. Numerous attempts to improve the customer process resulted in negligible changes, such as stylistic changes to the document they need to fill in. The cumbersome process of filling in a complex document by hand, and contacting Luminus several times, remained basically the same. Now, as a result of the transformation, all dealings for the customer are handled by a one-call-to-move".

So, what's the secret behind these improvement projects that make life easier for both customers and Luminus? Henri Buenen: "Always having that long-term goal of customer orientation in mind is vital. A sharply defined end-vision should be put forward for each project. This keeps you from deviating from core objectives when discussing internal implementation". Nele Aendekerk adds: "Sponsorship and ownership by the business – from launch through completion – is crucial. Also, as of the initiation phase, buy-in from people cross-departmentally is required. And then even in the following stages, development and implementation should be guided by a cross-departmental group of motivated employees".

Process + People = Customer Experience

Another ingredient is explicit project descriptions. Henri Buenen: "Customer orientation must not be just a slogan. For each project, the impact on every department involved – and what will change in the daily way of working – must be made clear. The objectives, the expected benefits for the customer and for Luminus, and the output related to process, people and system should be outlined".

But the transformation has not only required technical change—Henri and Nele believe that effecting a cultural change has been equally important. Nele Aendekerk: "Luminus is trying to stimulate a healthy sense of individual responsibility in each employee. Each individual should grasp how his or her job and behaviour contribute to customer orientation. For customer-facing contact agents, this has entailed measuring customer satisfaction on an individual level. Non-customer-oriented behaviour can be exposed this way, but the main purpose is to make the customer experience transparent at different touch points to discover where underlying processes might be underperforming. Low scores might mean that the agent didn't have the resources or the authorisation to help the customer efficiently".

Henri believes that a fine balance exists between a good customer process and friendly, customer-oriented employees: "Friendly people do not make up for a bad process, but a good process cannot run without friendly people". Nele agrees: "The performance of front-line agents will be only as good as the underlying processes. Coaching and meaningful incentives are essential. It has been crucial to give non-customer-facing employees an end-to-end view so that they can relate their personal job to the chain of activities that support the customer process".

Regarding New as Normal

Even taking all of these things into consideration, Nele acknowledges that: "There will always be a substantial lag between the implementation of a process change and the moment employees regard this changed process as the daily way of working, the normal way". Employees need a lot of guidance for 'the new process' to become 'the process'. And this is something that is easy to underestimate. Having too much happening at once can jeopardise acceptance of change—you risk ending up at 80 % in all projects. Exploring the impact of a new process by means of a well-tested pilot is one way to promote acceptance; but Luminus has found that scaling up to the entire organisation introduces a new set of challenges and requires advanced change management.

Looking ahead, Luminus foresees opportunities in e-care (social media) and in intelligent information at the client level (intelligent meters, energy producer in your own house,...). But before these opportunities can be pursued, Luminus is determined to 'focus on finish': taking the transformation one step at a time and generating tangible results that all contribute clearly to a customer-oriented organisation.

Henri Buenen is Director Customer Service at EDF Luminus, where he is responsible for the commercial and operational results of the customer service department. His achievements in this role include developing the company customer experience strategy and implementing the new customer-oriented service organisation. Working at Luminus for more than 10 years, Henri has been B2B Director and Marketing Manager. Prior to Luminus, Henri gained experience in the energy industry at Essent.

Nele Aendekerk is Customer Experience Manager at EDF Luminus, where she is currently co-driving the implementation of a more customer-oriented service organisation. Before Nele devoted herself to creating a better customer experience, she performed various roles at Luminus including Segment Manager in the professional market, Quality and Knowledge Manager, and Process Manager. Nele's expertise lies in customer experience, project management, process and business improvements, product management and segment management.

BPM Meets Social Software: An Interview with Hajo Reijers

For years now, BPM technologies have enabled organisations to work in a process-oriented manner. But the recent invasion of social software tools has not gone unnoticed in the BPM arena, and so the BPM technology landscape is undergoing profound change. Prof Hajo Reijers from TU/Eindhoven discusses the opportunities that arise when BPM practitioners leverage social software technologies.

Prof Hajo Reijers specialises in empirical research on technology-oriented (re)design of business processes. Yet, this technological focus does not exclude the organisational component from his conception of BPM. Hajo Reijers: "Business processes are an organisation's most valuable assets – they're the core of the organisation. Companies can leverage these assets in various ways, including measuring and evaluating process orientation, improving the processes, fostering process ownership and governance, and so on. Technology can play a vital role as enabler, but it's only a means to an end."

No matter how straightforward managerial principles of horizontal workflow management may sound, organisations struggle to grasp what this can mean for their business. Hajo acknowledges that "process management may sound to some as vague and intangible, while the concept of 'workflow management' may be hard to fathom too." He adds that, on the other hand, there are organizations that overly focus on the technological side of BPM. "It happens that BPM suites have been used successfully to shed light on the opportunities and benefits of BPM, and are henceforth portrayed as the *sole* embodiment of BPM."

Overall, BPM tools have not become mainstream, and vendors have not witnessed a spectacular expansion like has happened with, for example, ERP tools. Prof Reijers sees a popular association between BPM on the one hand and, organizational inflexibility and over-standardisation on the other. "BPM tools are deployed in many large organisations, yet these applications are often implemented in a 'monolithic' way. You install it, you keep away." And this is a wrong approach, which forgoes numerous opportunities for BPM and BPM technologies.

J. Van den Bergh et al., *Transforming Through Processes*, SpringerBriefs in Business Process Management, DOI: 10.1007/978-3-319-03937-4_8, © The Author(s) 2014

Enter Social Software

The basic principles of social software can be applied to unleash the inherent power of BPM technologies. "Social software is one of those technologies that organisations can leverage to manage complex and unstructured problems." Ad hoc communication, external to the organisation, and case-by-case, are typical characteristics of social software. Creatively employing these principles allows companies to distribute work in a completely novel manner.

> Consider Twitter: this online social networking and micro-blogging service collects huge amounts of data, and it's up to users to make sense of these data – for example, in better understanding customer demands. Similarly, employers can make use of the global workforce that's available through social software. Companies can expand their skill set significantly by accessing highly specialised human skills worldwide and outsourcing complex and case-specific tasks.

Global Online Workforce

Organisations are only on the verge of taking advantage of ubiquitous social technologies like Twitter and Facebook, but Hajo sees 'crowd-sourcing' principles quickly emerging. Take the example of Amazon's 'Mechanical Turk' (http://www.mturk.com), where businesses and developers pay an on-demand online workforce to perform small 'human intelligence tasks'.

Other examples can be found in the healthcare industry. Standard procedures dictate that certain X-ray scans must be read by two different radiologists. Conventionally, both radiologists work in the hospital in which the scan is performed. To speed up this procedure, to lower costs, and to benefit from radiologist specialisations, radiology departments are starting to make use of external radiologists to perform the second-opinion check. Even more, a new industry of 'second-opinion radiologists' has sprouted up.

These are but two examples of how companies are leveraging knowledge and labour that are available throughout the world.

New Generation of BPM Tools

Hajo expects this phenomenon to become integrated in a new generation of BPM tools. "The job of digging in data for relevant information and automatically collecting significant information from various sources can be supported by technological tools. As for organisational skills and capabilities, businesses need to become smarter in recognising patterns. And this is where BPM experts can learn from marketing. For years, marketing has developed expertise in techniques for discovering patterns in customer data. Doing the same for the internal organisation will require capabilities similar to those found in marketing departments."

On the other hand, organisations must overcome some practical complications. "Organisations are reluctant to embrace social software, because they're concerned about giving up control over their internal processes. Finding that balance between added value and practical and legal considerations is still a hurdle. But social software has already been adopted by so many individuals – organisations simply can not continue lagging behind. Social software technologies are the next-generation means for improving business processes and tackling special, complex organisational problems."

Hajo Reijers is a full Professor in Computer Science at the Technische Universiteit Eindhoven (TU/e), where he holds the Chair in Business Process Technologies. In combination with his professorship, he leads the BPM Research group at Perceptive Software. Furthermore, he is affiliated with the TiasNimbas Business School. He is also one of the founders of the Business Process Management Forum, a Dutch platform for the development and exchange of BPM knowledge between industry and academia. Earlier in his career, as a consultant for Deloitte and Accenture, Hajo gained ample hands-on BPM experience in a variety of reengineering projects and workflow system implementations.

BPM in a BRIC Country: Spotlight on Brazil—An Interview with Professors Marcos Paulo Valadares de Oliveira and Marcelo Bronzo Ladeira

In today's globalised environment, it's easy to assume that BPM practices are similar worldwide. Yet, institutionalising BPM is influenced strongly by local market characteristics and geographic and corporate cultures. So, we investigated BPM practices in developing BRIC countries—and specifically, Brazil. We asked Brazilian academics Prof Dr Marcos Paulo Valadares de Oliveira and Prof Dr Marcelo Bronzo Ladeira (Universidade Federal de Minas Gerais, Brazil) how BPM is different in Brazil, a country with a distinctive business culture.

Brazilian companies do resemble Western companies in terms of BPM mindset: for many years, managers in both foreign and Brazilian companies in Brazil have been talking about BPM but not really acting on it. The majority of them focus on trying to break intra-organisational barriers to supply chain integration, thus missing out on some of BPM's more strategic virtues.

Marcelo Bronzo Ladeira: "Most managers in Brazil do not regard BPM as an opportunity to institutionalise end-to-end process thinking and continuous improvement to generate customer value. Instead, BPM is regarded as a tool for streamlining processes and reducing costs without concerning the customer directly. Activities are focused almost exclusively on operational aspects, and they're often even restricted to process mapping and documentation."

Yet, both professors see a movement towards BPM as a management discipline, supported by an increasing demand for formal BPM training. In some cases, practices related to total quality management, lean or Six Sigma are setting the stage for further development of BPM.

Challenging Country Characteristics

Despite similarities in BPM maturity compared to Western countries, Professors Valadares de Oliveira and Bronzo Ladeira_point to the fact that the BRIC countries are dealing with some particular challenges. For Brazil specifically, a first dynamic affecting BPM maturity relates to market structure. Brazil is almost continental in terms of size, with 99.3 % of the Brazilian organisations classified as micro and small-sized companies. Research shows that, in comparison with large companies, BPM maturity levels are relatively lower in small- and medium-sized companies, which often focus on rapid growth and aspire to grow organically without installing formal structures and procedures. Management mechanisms such as BPM are frequently viewed as growth inhibitors—a perception that actually hampers Brazilian companies, because efforts to coordinate and manage supply chains demand inter-organisational process integration.

Furthermore, market characteristics (e.g. high tax load that restricts the number of global business transactions) and a logistical infrastructure that is insufficient for Brazil's geographical size present numerous structural challenges. Still, Brazilian companies demonstrate vitality and resilience: they usually have the capacity to cope dynamically with unexpected changes in order to remain competitive, grow, and exploit new market segments.

Another crucial characteristic that affects how Brazilian companies implement BPM originates from corporate culture. In general, corporate culture in Brazil is less formal than Western or Asian corporate cultures, as Brazilian workers respond less positively to strict procedures and hierarchy. This phenomenon must be managed carefully, because it can promote flexibility and process innovation. On the other hand, informality on the work floor can also cause chaos and complicate efforts to implement formal work procedures. So, organisations need to find a balance. To this end, all BPM implementation efforts must include change management as a critical success factor to avoid stifling the innovation that results from a flexible culture and losing control due to a lack of formality.

Standardisation Versus Informality and Flexibility

Due to these country-specific features, Brazil is characterised by uncertainty and informality. This creates difficulties, especially for companies trying to standardise processes in order to benefit from cost economies and produce on a large scale. Marcelo Bronzo Ladeira: "Having to deal with high uncertainties jeopardises standardised process management and continuous improvement. On the other hand, companies facing less uncertainty are often more mature in BPM."

Prof Valadares de Oliveira illustrates this with the example of public institutions in Brazil: "We find that BPM success stories are more prevalent in public institutions than in private companies. Pressured to take a more professional approach – as opposed to being politically driven – these institutions are actively engaged in BPM to attain higher efficiency and effectiveness. They're also less vulnerable to external risks and uncertainties. Market characteristics and infrastructure (such as sudden inflation, road blockages, etc.) have only minor impact on their daily functioning. Therefore, their BPM efforts are more likely to be effective and sustainable."

Yet both professors agree that process standardisation in Brazilian companies must be handled prudently: "Process standardisation is an important condition for reaching effective management results. It can deliver significant improvements in cost and/or quality of goods and services, which enhance the company's competitive position in both domestic and global markets. However, rash standardisation might kill the spontaneous innovation that results from a flexible work culture."

BPM as an Opportunity

So, does BPM in the developing BRIC countries differ much from BPM in Western countries? Not according to Marcos and Marcelo: "Despite particular market, cultural and logistical characteristics, Brazilian companies face the same problems that all other companies face when it comes to understanding, and reaping true benefits from, BPM. Managers all over the world must look at BPM as an opportunity to change their company's way of thinking and to continuously improve processes, instead of restricting BPM to documenting process flows."

Marcos Paulo Valadares de Oliveira is Full Professor in the department of administration at the Federal University of Espírito Santo, Brazil. Marcos has published in several local and international books, periodicals, and conferences such as Decision Support Systems, Supply Chain Management: An International Journal, IPSERA, EUROMA, POMS. His research interests include supply chain maturity models, supply chain risk management, business process Management and analytics.

Marcelo Bronzo Ladeira is Full Professor at Federal University of Minas Gerais—Brasil. His main research interests are focused on logistics and operations performance; best practices of supply, manufacturing and physical distribution in the automobile industry; BPO and SCM processes maturity.

A Cultural Fit for BPM: An Interview with Alec Sharp

Despite his expertise in process management, Alec Sharp's career has not been a carefully planned and executed process. He would rather describe it as 'a series of happy accidents' that has brought him to where he is today: a highly respected, popular facilitator of business change.

Alec Sharp serves corporate clients worldwide with his remarkable adaptation skills and contagious enthusiasm for process management. Those who have had the pleasure of seeing him give one of his many invited lectures will concur with our appreciation for his expertise. We wanted to hear about the insights into BPM and culture that he has gained over the course of his career.

Alec Sharp: "Organisational culture is mainly observable through individual and organisational behaviour (what people consider to be good behaviour), which is driven by values and beliefs. The McKinsey definition of culture—'how things get done around here'—works perfectly for me. An organisation's culture can be instilled explicitly or implicitly, but mostly it's a process of osmosis in which people gradually pick up acceptable behaviours by observing the actions of others, and the positive or negative consequences."

> This can be the sum of local subcultures, team- or geography-based, but usually there is an overriding culture. However, I have witnessed global organisations have significant difficulties when they impose processes that do not fit the local culture – so it is certainly an important factor. In one case, a business process that was designed in the United States was unsuccessful in a southeast Asian country until it was revised to reflect that country's national culture and business norms.

Changing Processes and Cultures

With regard to BPM and culture, Alec does not believe there are organisational cultures that are favourable or unfavourable towards BPM. Organisations are simply trying to get things done, the way they believe they should be done, regardless of sector or size. "Sometimes, when process orientation doesn't take

J. Van den Bergh et al., *Transforming Through Processes*, SpringerBriefs in Business Process Management, DOI: 10.1007/978-3-319-03937-4_10, © The Author(s) 2014

hold, it isn't a matter of culture – it's just that the word *process* puts some people and organisations off. They've had a history of 'business process reengineering', and sometimes the word 'process' has been compromised and misinterpreted. In those cases, I use the term 'business change' as an alternative."

Alec suspects that a sub-current in the BPM field may be responsible for that aversion to the word *process*: some BPM practitioners are too occupied with creating labels, standards, centres for excellence and dogmas instead of recognising that the essence of business change is human, organisational and cultural. So, if BPM fails, it's not a failure of the BPM discipline as such, but rather a failure of the BPM approach taken by the organisation within the existing culture.

If culture plays such a prominent role in BPM success—or business change in general—should we not try to change an organisation's culture to create a cultural fit for BPM? "Well, it's certainly not a good idea for a BPM group, process manager, or centre for excellence to set out to change the organisational culture to kick off their project. That's a recipe for a very frustrating time. Changing an organisational culture is virtually impossible unless the change process is concerted, long-term and top-down. It often makes much more sense to change the company's operations first, which leads to a bottom-up culture change. I've encountered several cases where process change has indeed led to cultural change. My advice is to play with the cards you've been dealt and start working on process in the current culture – it's possible!"

Start with Small Successes

Obviously, there are reasons for BPM failures that go beyond culture. "First of all: failing to identify business processes end to end. This happens most frequently in organisations that have adopted a very technocratic (possibly Six Sigma inspired) approach that focuses on specific activities rather than on the entire end-to-end business process. If a company applies these techniques to a single activity, without understanding the impact on the end-to-end process, it's liable to just make things worse."

In addition, there's always the danger of not paying attention to human enablers, such as motivation and measurement, skills, and roles and responsibilities. Designing the perfect process has to go beyond the obvious factors to areas that are, perhaps, less comfortable for process practitioners. "It's possible to design the perfect process in a process diagram and still miss the target by miles if you fail to take on a holistic view with attention to human and cultural factors."

As difficult as implementing BPM is, we tend to make it even more difficult on ourselves by trying to tackle every problem at once. "We're often too impatient in our field, precisely because process is such a powerful lens for looking at an organisation's functioning." In his own practice, Alec prefers to start with a more incremental approach. "Start out with building small successes, and gradually spread the word by celebrating those initial successes." Although 'top management support' is one of the most mentioned success factors in the literature, Alec is convinced that most top managers do not think in terms of business processes, and so they will not likely initiate process orientation. "Still, I do see cases where BPM, through an incremental process, finds its way up to the top and results in an enterprise-wide approach. So, don't wait for top management support to initiate some action on the process side."

Credibility is Key

Based on his many observations of success, Alec specifically promotes consistency and repetition: "Don't constantly readjust your message based on the fad or flavour of the day. Working with simple tools, and transmitting the same message repeatedly, greatly increases your chances of success. Practitioners should not take refuge in advanced toolkits if the organisation is not ready. 'Credibility' is the key word if you are looking for a suitable sponsor to provide leadership and support for your process management initiatives. The ideal sponsor is a good experienced manager – and, preferably, a good communicator too – with high credibility."

Finally, a word of caution from Alec to BPM practitioners: "I have observed success and failure throughout my career, and far too many people in the BPM field seem to think that their particular view on business processes will make sense to everybody. They do not make sure that everyone has the same understanding of what a 'business process' is in the first place, and they do not take organisational culture into account."

Alec Sharp is a consultant based in Vancouver and sought by companies worldwide that are in need of guidance on process management and requirements modelling. He is the author of the best-selling book Workflow Modelling (Second Edition) and a regular guest speaker on BPM topics in workshops and events on all continents.

Focusing on BPM's Human Factor:
An Interview with Els Van Keymeulen
of Schoenen Torfs

Considering the fact that, in most processes, it's still the human factor that makes all the difference between success and failure, a booklet on BPM certainly needs to include an SME's particular insights, and most certainly it needs the employee perspective. Els Van Keymeulen, HR Manager at Schoenen Torfs, embodies the belief that happy employees create happy customers.

Els Van Keymeulen: "Our employees make all the difference! That's a strong statement, and all too often proclaimed – but you will seldom find a place where it is as obviously true as in the Torfs's family-owned shoe business, Schoenen Torfs." When you enter the company's head office, the walls covered with 'Great-place-to-work' awards tell the story. We asked Els about the business logic behind the awards and how employees and processes are complementary in this medium-sized business.

Caring for Employees

Els Van Keymeulen has been the company's HR Manager for 17 years now, supporting CEO Wouter Torfs in stimulating employees to deliver excellent customer service day-in day-out. "The passion for our products and our business starts here. At Schoenen Torfs, all managers are expected to be role models in being passionate about their jobs, caring for colleagues and striving for excellent customer service. Although the family business has grown to become a company of over 500 employees, its DNA still radiates a large family feeling."

Despite such impressive growth, complexity is being kept to a minimum and the organisational structure is still fairly flat. As collaboration is essentially part of the company culture, Schoenen Torfs does not suffer from isolated silos and turf protection, and the company does not feel the need for formal process roles and process management initiatives.

Caring for employees has a prominent place in the company's vision. Schoenen Torfs's focus is all about the belief that happy employees will drive better sales

J. Van den Bergh et al., *Transforming Through Processes*, SpringerBriefs in Business Process Management, DOI: 10.1007/978-3-319-03937-4_11, © The Author(s) 2014

results when they convey their enthusiasm to customers. Their core business process—the shopping experience—is analysed on a regular basis by means of mystery shopping, market research and company image reports.

"Our CEO is constantly pushing his management team to see the broader picture and disseminate our vision. And the employees can give input to improve our business, although we could probably involve them even more." This feature of Torfs' strategy has been present for generations—the grandparents of the current generation were already renowned for their caring nature and continuous attention to excellent customer service.

Inspiring Customer-Centricity

You do not grow a company culture overnight—and Schoenen Torfs has been working on its employee-centric approach for many years, deploying a number of initiatives to that end. "Torfs's success has brought growth as well as more organisational complexity. The role of the middle managers, on the regional and shop levels, has been a point of attention for many years. They have been carefully selected, and we have been passing on our passion to them. Moreover, we have worked intensively with them on leadership, customer-centricity and working constructively with employees."

To engage all employees and spread the message, Torfs's management team employs role playing, workshops, semi-annual employee days, etc. "We spend a lot of time thinking about new workshop formulas. It needs to be fun above all, as we strive to create a community feeling among our employees. Routine is the enemy, stimulating us to find new ways to engage our employees again and again."

Over the years, the company and its leadership have become more professional in their approach. One of the exercises led to the definition of a set of core values—but, unlike many other corporate values, these core values resulted from a collective employee effort. As for the recruitment and selection process, it is strongly value-driven: a good fit with the company values is a major criterion.

Balancing Standards and Flexibility

Managing processes is also striking a balance between standardisation and flexibility. "In our company, we tend to favour flexibility over procedures. Not getting lost in the details that are hardly relevant to the end result does the trick. That doesn't mean Schoenen Torfs is losing sight of efficiency: it means that the workforce's energy should go into sales, serving the customer, and creating a long-term perspective for the company."

Nevertheless, company growth has brought an increasing need for standardisation, driving Torfs's management team to consider where standards and procedures are essential. When procedures are deemed necessary, Els says they look for transparency and simplicity and a strong link with the organisation's goals. Even the performance evaluation process, for example, is kept to a bare minimum. The assumption is that managers at all levels should coach, steer and provide feedback on-the-job throughout the year. Evaluation cannot be reduced to a single moment in time.

On the other end of the spectrum are the suppliers. Managing its processes beyond the company walls, Schoenen Torfs partners with suppliers and promotes win–win solutions to create a sustainable relationship. This approach has resulted in process optimisation from end to end.

Long-Term Competitive Advantage

A key element in the company's ambition to be number one in customer service is rigorous and regular measurement of what the customer service level is and then acting on that information. With a consistent picture of growth and success, Schoenen Torfs seems to be well on track. In turn, this fortifies their position on employee-centric management, which is regarded as a catalyst for long-term competitive advantage.

"Make no mistake: our focus on employees doesn't mean we've lost sight of the end-result. We want to be successful." Personally, Els Van Keymeulen maintains her drive—even after 17 years—by constantly finding new ways to stimulate employees and combat routine. "We're constantly looking for the best process to convey our vision and strategy to our people."

Els Van Keymeulen is a passionate HR professional with over 17 years of experience in managing the human capital of a medium-sized enterprise. Her work at Schoenen Torfs—combined with the many efforts of the CEO and his team—has been recognised by multiple Great-Place-to-Work awards for the 'company of 500+ employees' category in Belgium. In 2013, Schoenen Torfs was the runner-up in that competition.

Mitigate, Reduce or Prevent: Managing Risks in Business Processes—An Interview with Kevin McCormack

Events over the past decade or so have shown the importance of understanding and managing risk in business processes, especially in those that have a global outreach. But recent crises have also revealed that many organisations lack effective risk management practices, despite increased attention to the topic. We asked Dr. Kevin McCormack, a leading BPM and business analytics consultant and author, how BPM and risk management can be related.

Dr McCormack first had his eyes opened to the pervasive nature of risk when he saw the disastrous impact the 9/11 attacks had on US businesses. Until then, major disruptions had been regarded as external factors beyond a company's control, which would require an adequate response if they occurred.

But, risk management includes much more than earthquakes, tsunamis, volcano eruptions and super storms. For example, new technologies can be as destructive to businesses as they are beneficial. Furthermore, globalisation, outsourcing and the Internet have created a more complex environment for business processes, especially on the supplier side.

In short, risks can be anywhere, both inside and outside a company. And if something is expected to have an impact on business processes, it's relevant to the company. Managing that risk proactively is exactly what the forward-looking manager should do.

Managing an Intangible

So how can organisations understand and proactively manage an intangible thing like *risk*? Kevin McCormack: "I use the ISO definition, which says that risk management is a systematic assessment of potential disruptions that may affect an organisation's ability to achieve its objectives."

> Then, the aim is to either reduce the probability of disruption taking place or to mitigate negative effects. The reliability of a company's business processes is a major issue, because customers expect a consistently reliable process output. Most companies are doing a reasonable job at managing internal risk, especially from a financial perspective. However, many are not up to speed in coping with external factors.

J. Van den Bergh et al., *Transforming Through Processes*, SpringerBriefs in Business Process Management, DOI: 10.1007/978-3-319-03937-4_12, © The Author(s) 2014

In his consulting and research activities, Kevin observes that risk management seldom receives the attention it deserves. "I see many companies discussing the topic, but all too often they have to admit that they're not really acting on it. I suspect that they're not really aware of the consequences, and put off by the foresight of investment. Those organisations are balancing on a very thin wire. I've seen too many cases where unexpected events could have been less impactful – if not actually prevented in the first place – with robust risk management."

Building Risk Management Practices into Business Processes

Nevertheless, Kevin has also witnessed some encouraging practices by pioneers in the risk management field. In particular, the aerospace and defence industries offer notable examples of risk management pioneering. Unsurprisingly they are often advanced when it comes to BPM practices too. In fact, they employ a process perspective to build risk management practices into their end-to-end business processes.

And that's precisely where Kevin sees BPM and risk management intersecting. "Although one is possible without the other, it's a tremendous advantage when organisations combine risk management with a process perspective and even use their process management capabilities to control risk. Process owners, in particular, are well placed to assess both internal and external process risk. They can play a 'risk management champion' role as a point of contact for the risk manager."

> In every organisation, the considerable responsibility for managing risk rests on every person's shoulders. Risk managers can act as lobbyists reporting risk, educating colleagues and installing standard risk management practices—but they can achieve only as much as their colleagues support.

Managing Risk from a Process Perspective

A number of success factors pertain to implementing risk management practices. "First of all, the terminology is completely alien to many managers. So, number one would be to educate the management team. Secondly (and this is a major pitfall), a company needs to be able to justify investing in risk management. Obviously, it's a challenge to measure the impact of uncertain events. Nevertheless, there are metrics that enable you to make certain estimations. Moreover, risk management from a process perspective will likely also improve business processes in terms of reliability and repeatability."

Kevin identifies the 'command and control culture' ('That is not my job') as a typical stumbling block. "Although many would agree that risk management is important, most would say that it's not part of their role. On the contrary, being successful in risk management requires the active contribution of all key stakeholders in the organisation. Again, BPM – by definition, the opposite of 'command and control' – can enable a company to decentralise the risk management responsibility."

Is there a zero-risk point, and is it possible to reach? "I think it's unlikely that such a point exists. As Forrest Gump says: 'Stuff happens, and it's going to keep on happening.' It's a matter of mitigating risk in so far as that is affordable. I definitely encourage organisations to use BPM and its supporting structure to manage process risks. A process-oriented company is undoubtedly better equipped to manage risk proactively. Horizontal thinking is a huge enabler to raising attention to – and acting on – risk management."

Dr Kevin McCormack, President of DRK Research and Consulting LLC in the USA, is a lifelong learning practitioner, author and lecturer in Business Process Management, Supply Chain Management and Business Analytics. His recent works have focused on risk management practices in businesses with global outreach.

The Next Wave in Process Thinking: An Interview with Hendrik Vanmaele

BPM followed more or less naturally in the footsteps of Business Process Reengineering (BPR)—and now we're witnessing the next wave in process thinking. This is Prof Hendrik Vanmaele's point of view, based on his observations and extensive experience in BPM projects. We asked him about what to expect next.

Hendrik Vanmaele—founder and CEO of Möbius, a consultancy firm specialised in Business Process Management and Supply Chain Management—comes from the academic world and based his start as an entrepreneur on his Supply Chain Management expertise. Hendrik Vanmaele: "In the 90s, production companies discovered the virtues of optimising their core operational processes and installing ERP systems. I soon noticed that service companies would benefit from similar principles and techniques. And by the early 2000s, BPM was born as the next logical step.

> But BPM back then was nothing like it is today. It was practically equal to mapping and modelling business processes in order to control them. Companies were investing massively in visualising processes in great detail. A second pervasive aspect of that early BPM was the focus on tools (which, naturally, was encouraged by the vendors). Dozens of tools – often with way too many features – were acquired. We all did it, and there was nothing wrong with that, except that there is so much more to BPM than tooling and flowcharting.

BPM's Heritage

Today, in contrast, BPM focuses primarily on project work and strategy implementation. "Projects are seen as a direct way to improve the performance of end-to-end processes, which is a vital condition for BPM to be accepted in the current economic turbulence. Strategy implementation has emerged only recently and it's still on the rise. BPM can potentially optimise the vertical process of cascading an organisational strategy through all levels of an organisation. So, it has become a general management capability."

J. Van den Bergh et al., *Transforming Through Processes*, SpringerBriefs in Business Process Management, DOI: 10.1007/978-3-319-03937-4_13, © The Author(s) 2014

Process modelling and BPM suites have not disappeared, but they have been put into perspective and are now more correctly viewed as tools and documentation. "One more aspect of 'the old BPM' was the eternal question: how to bring BPM to the board level? Today, the preaching part is behind us – those who haven't seen the light will inevitably stay behind, but most boards have bought into BPM anyway. Whether their actions always follow their beliefs is another matter."

Hendrik thinks that the BPM acronym will probably not last much longer. "But the process perspective, which existed long before BPM was called BPM, will definitely continue. Call it BPM's heritage: putting activities in a larger end-to-end perspective that transcends departmental and organisational boundaries. True to its holistic nature, BPM has been converging with change management, IT management, customer orientation and quality management to become what it is today. BPM as a domain is slowly fading – it won't disappear completely, but it will reincarnate as part of a larger entity."

Focusing Attention on the Process Perspective

How is all this impacting the process support organisation? "BPM and similar performance improvement initiatives have led to a proliferation of staff functions – think of business process offices, quality departments, audit teams, project offices, and so on – all working towards the same purpose with a lot of overlaps. Like the disciplines behind them, these staff functions are converging into a single office (call it a programme office or board), which bundles analytical expertise, focuses initiatives and coordinates all strategic efforts to improve organisational performance."

What would BPM's 10-year progress report be? Has BPM accomplished what it promised? Or should we expect, like so many Business Process Reengineering reports at the end of the 90s, criticisms that BPM has failed to deliver better performance? "I don't think BPM will end up like BPR. Despite over-investment in process modelling and heavy tools, I think BPM has largely lived up to its promise. I have seen plenty of good examples, in which BPM has succeeded in focusing attention on the process perspective to the benefit of organisational performance. Few people will still claim that a classical bureaucratic organisation functions better than a process-oriented organisation."

A downside to our BPM efforts, certainly those of a few years back, is that the customer has too often been omitted. BPM definitions usually define the end-goal as 'value creation for the customer' - yet, most process improvement efforts have been performed from an inside-out perspective. At best, the customer has been regarded as a stakeholder in the process.

Large corporations, particularly in the services and utilities industries, have gone through a trust crisis with their customer base. 'New BPM' should help resolve that crisis, but only if companies adopt the customer perspective. Success stories, like the one at Carglass,[1] may inspire other organisations to strive to understand customer needs and install efficient processes based on that knowledge.

Mastering Processes to Deal with Complexity

To master the 'new BPM', organisations will have to develop, or acquire, certain capabilities. It's not a matter of having the right software tools. They need to possess essential intangible assets, such as: management momentum (the ability to turn strategy into action), modularity (which promotes organisational agility), and customer experience design (the ability to understand the customer and integrate the customer into processes). Finally, change management (successfully guiding change) remains a key capability to make the other assets work.

Meanwhile, the need for process awareness is still increasing. Continued globalisation, a challenging economic outlook, loss of customer trust, etc. are forcing us to keep investing in processes and how they support business models. On top of that, new technologies are constantly shaking up the business landscape. I've seen a lot of uncertainty in management teams. Non-IT-savvy managers, in particular, sometimes fear the surge of new technologies. More than ever, they expect a proof of concept.

Nevertheless, Hendrik believes that technology will once more be a key driver for process innovation. "BPM has a future, albeit probably under a different name, in mastering processes to deal with complexity, implement strategy and serve customer needs."

Starting from a focus on Supply Chain Management at Ghent University, Prof. Hendrik Vanmaele has spent most of his career as an academic and as a consultant in the fascinating world of Business Process Management. In the late 90s, as a spin-off from the university, he founded Möbius, a research and consulting company specialised in BPM and Supply Chain Management. Over the past decade, Möbius has evolved into a strong player in the strategy implementation consultancy market.

---

[1] Carglass is the Belgian subsidiary of the international Belron group, specialist in car window repairs and replacements. The company has been extremely successful with a customer delight campaign, supported by a strong process focus.

Ambidextrous BPM: Making BPM Exciting Again—An Interview with Prof. Michael Rosemann

Prof Michael Rosemann is not surprised to see BPM on the list of management disciplines that are going through the so-called trough of disillusionment. For too long now, its practitioners have used BPM only to exploit existing business processes. It's time to start exploring too—and make BPM exciting again!

Following the lineage of Business Process Reengineering, Systems Thinking and Quality Management, BPM has come of age as a management discipline. Deeply rooted in the continuous improvement movement, BPM has proven its value for improving business processes. But does BPM have what it takes to drive innovation and breakthrough thinking as well?

Michael Rosemann: "We know how to model, analyse, automate and streamline processes. We understand how BPM can help us to overcome identified problems within a process. In fact, the related capabilities can nowadays be regarded as a commodity. But, so far, BPM as a discipline has not been good at supporting and driving 'opportunity-driven' – instead of problem-driven – initiatives. I believe that this must be the next area of interest for the BPM discipline to remain relevant within organizations looking for innovative approaches."

Exploitation and Exploration

Business 'ambidexterity' is the balancing act between exploitation and exploration: the former focuses on ensuring transactional excellence with a concentration on net cost reduction, the latter is centred on transformational excellence targeting net revenue generation. And organisations need to be able to do both at the same time, even if they require different capabilities.

Process exploitation is about inside-out, reactive, problem-driven process management. Exploitation examines a process and fixes what is broken. I call this an organisational hygiene factor: it doesn't excite customers or managers, but it's a precondition for staying in business. Process exploration, on the other hand, is often driven by outside opportunities (e.g., emerging technologies) and is proactive. It has the potential to deliver tremendous added value, and high levels of customer delight, by offering new services or innovatively transforming existing services. And it poses a new and exciting challenge for BPM practitioners and researchers. However, while current BPM methods, tools and techniques support process exploitation well, process exploration is still in its infancy.

As an example, consider designing a new process for the annual lodgement of income taxes. A classic BPM approach to improving this process would be to analyse the process, look for the shortcomings (e.g., bottlenecks, rework), and then reactively improve these broken parts of the process. Contrast this with a 'process explorer', who looks at available technologies, established practices in other industries and takes an outside-in perspective on taxation. He or she might envision real-time taxation just like 'real-time insurance' and then work from there to fundamentally rethink the process and related services. The process explorer works backwards from a strategy-driven (and hopefully exciting) vision of the future, while process exploitation is focused on the (frustrating) current weaknesses of a process. Today, this type of breakthrough thinking usually occurs without involvement of the business process office.

Take new exciting technologies such as the Internet of Things or social media – BPM professionals need to complement their existing toolkit with approaches that help them to translate such emerging new opportunities into entire new process experiences. This requires a shift in thinking from 'pain points' to 'opportunity points'.

The Outside-in Perspective

To help usher in this next phase of BPM, Michael sees Customer Process Management (CPM) as a way of tapping into private processes with value-adding services. The better these services blend in with a customer's private processes, the better they will be received. In other words, the birth-to-death-value chain as the ultimate business process and the ultimate form of customer-centred process design.

Exploration and outside-in thinking promote both looking at the entire experience of going through a process rather than merely at the detail of the process or the process outcome. Take my own organisation, Queensland University of Technology: in the past, our customers' expectation was to come to the campus, pay for a course, sit in classes, do assignments, an exam, and (ultimately) graduate. Now, using process experience thinking, the challenge is not to improve in-class lectures, but to create a new service experience, in which, by means of mobile and context-independent learning, students take micro-courses when and where they want and pay incremental fees ('pay as you learn').

Embedding Ambidextrous BPM

A process explorer could be characterised as a highly extroverted individual—a future thinker, capable of crafting a process vision detached from any obstacles or practical objections, and finally turning this vision into reality. In a way, he/she translates (technological) opportunity into (new) process design. To make all that work, process explorers need to be able to interact with all kinds of people in the organisation.

Michael believes that most business process offices in organizations are currently populated by more analytical, inside-out thinking profiles due to the traditional focus of BPM on strong modelling and analytical capabilities. So it's very likely that process exploration needs to be performed by other people, and presumably the organisation will need other tools too.

> The ambidextrous organisation tells us that you need dedicated people with the right skill set for the process exploration job. Process innovation and exploration already exist, but they're usually unstructured and they hardly ever come from a BPM team's initiative. Nevertheless, both process exploitation and exploration deserve a place in the business process office's service portfolio. To truly embed ambidextrous BPM, business process offices should attract the right profiles and make exploration part of their service offering to the business.

Striking a Balance Between Exploration and Exploitation

So, how to devote optimal attention to both process exploitation and exploration? "I believe that's strictly strategy-driven. Those organisations that get their competitive advantage from cost-effectiveness, efficiency, etc. will probably lean more towards exploitation. On the other hand, the organisation that is losing market share will need to look more towards exploration, as they have to find new ways to shake up the market by identifying unexplored customer needs and creating new services."

Moving BPM out of the trough of disillusionment will be a major challenge for all BPM practitioners in the coming era of process thinking and business transformation. But at least we know it will be an exciting journey—with ambidextrous BPM at the centre of the movement!

Professor Michael Rosemann is Head of the Information Systems School, Science and Engineering Faculty, Queensland University of Technology, Brisbane, Australia. He is renowned for his expertise in Business Process Management and leadership in the BPM research arena with the BPM discipline group at QUT. He has published several books, co-authored more than 200 papers and delivered invited keynotes at all major BPM conferences in the world.

Acknowledgments We cannot end this work without thanking the contributors to the interviews. We sincerely enjoyed each of the conversations, and welcomed the open and enriching answers and insights of all of you. Furthermore we like to extend the acknowledgements to the member organisations of our Centre for Excellence (www.vlerick.com/bpm) over the past 9 years. Not least because of their continuous drive to become better, to materialise business transformation through process management. In alphabetical order we appreciate the exchange with Antwerpse Waterwerken, Belfius, Belgacom, bpost, Brussels Airport, Delhaize, Eandis, EDF Luminus, Elia, FOD Financiën, Jessa Ziekenhuis, MasterCard, OPZ Geel, Partena Ziekenfonds and Partners, Pioneer, RIZIV/INAMI, SaraLee, Stad Gent, VDAB, and VMM.